THE SECRETS OF SATOSHI

Understanding Bitcoin

Christopher Perceptions

CMPGFB LLC.

To my wife and children.

Bitcoin is a gift from God and an opportunity for all to have a chance at being part of a fair system on Earth.

CHRISTOPHER PERCEPTIONS

CONTENTS

THE BITCOIN WHITEPAPER

Under the alias Satoshi Nakamoto, an unidentified person or group released the fundamental Bitcoin whitepaper in 2008. The whitepaper describes the conceptual framework and technical underpinnings of Bitcoin and is titled "Bitcoin: A Peer-to-Peer Electronic Currency System."

The whitepaper by Satoshi Nakamoto describes Bitcoin as a decentralized virtual currency that enables peer-to-peer trades without the use of middlemen like banks. The term "blockchain" is introduced, which is defined as a public ledger that securely and openly records all Bitcoin transactions.

The document outlines how Bitcoin uses cryptographic proof in place of a trusted third party to address the long-standing issue of double-spending in digital money systems. It describes the procedure for transaction validation using the proof-of-work consensus mechanism, which entails miners competing to solve mathematical problems to validate transactions and safeguard the network.

Also, Satoshi Nakamoto outlines the decentralized nature of the network, the creation of new Bitcoins through mining rewards, and the advantages of anonymity and privacy in Bitcoin transactions. The main ideas, techniques, and incentives supporting the Bitcoin system are high-level summarized in the whitepaper.

The Bitcoin whitepaper was released and since then it has acted as a guiding principle for the growth of the Bitcoin network and served as an inspiration for the development of countless other cryptocurrencies and blockchain-based projects. It continues to be an essential resource for comprehending the underlying ideas and objectives of Bitcoin as a revolutionary financial technology.

INTRODUCTION

The first and most well-known cryptocurrency, Bitcoin, has drawn interest from people all around the world. You probably have a lot of questions regarding the nature, functionality, and potential applications of Bitcoin as a beginner to the space. We will cover 100 frequently asked questions about Bitcoin from beginners in this comprehensive guide. We will give you useful insights into the world of Bitcoin, from comprehending its basic ideas to investigating its real-life applications. Let's explore the secrets around this ground-breaking digital currency now. This guide will provide you the skills necessary to navigate the fascinating world of Bitcoin, whether you're interested in learning about its history, the complexities of its technology, or how it affects different facets of our life. My Bitcoin journey started in 2016. I read the whitepaper and it changed my life because for the first time, I could envision a fair system. I knew two things after reading the whitepaper.

1. Bitcoin was going to change the world.

2. In order for mass adoption to take place, there needed to be frictionless onramps and off-ramps. It starts with education.

This book was created to act as a compass for you as you learn about Bitcoin. In order to make Bitcoin adoption a living reality, I have educated thousands of people on a global, national, and local level through courses, speeches, personal interactions, as well as the development of products and services as a founder building on Bitcoin. Simply stated, Bitcoin has changed my life and I believe it can change the world. If you are reading this book, you are either curious, wish to be educated, or you want to see Bitcoin change the world as well. No matter what brought you to my book, I welcome you.

WHAT IS BITCOIN?

Bitcoin is a peer-to-peer network-based decentralized digital currency. Under the pseudonym Satoshi Nakamoto, it was developed in 2008 by an unidentified person or group. Bitcoin is not governed by a single entity, unlike conventional currencies that are issued by governments. To keep an accurate and secure record of all transactions, it uses a technology known as blockchain. Powerful computers solve challenging mathematical puzzles to create Bitcoins in a process known as mining. Bitcoin is an alternative type of currency that has the potential to upend established financial systems since it has the ability for quick and inexpensive transactions, worldwide accessibility, and a finite supply.

HOW DOES BITCOIN WORK?

Blockchain technology, a decentralized, open-source ledger that stores all Bitcoin transactions, is the foundation of how Bitcoin operates. A Bitcoin transaction is started by a user and broadcast to the network, where it is verified by miners and added to a block. A method known as proof-of-work is used by miners to compete to solve challenging mathematical challenges. The transaction becomes irreversible and unchangeable as soon as a block is added to the network. To provide security and avoid fraud, Bitcoin transactions are validated by a number of network users. There are also only 21 million Bitcoins available, and as time goes on, the rate at which new ones are created through mining will also decrease.

WHO CREATED BITCOIN?

Under the pseudonym Satoshi Nakamoto, a person or group designed Bitcoin. Numerous theories and suppositions have developed throughout time regarding the real identity of Satoshi Nakamoto, which is still a mystery. The Bitcoin whitepaper, which outlined the fundamental ideas and workings of the currency, was published by Nakamoto in 2008. Nakamoto was a key figure in the early creation of the Bitcoin program. Nakamoto disappeared from the public spotlight in 2010 and hasn't been seen or heard from since. The identity of Nakamoto remains a mystery, which has only increased interest in and fascination with Bitcoin. On January 12, 2009, Satoshi Nakamoto sent Hal Finney, a computer programmer, the first Bitcoin transaction. A proof of concept for the Bitcoin network, the transaction involved 10 Bitcoins, which at the time had no recognized market value. The exchange took place over the Bitcoin network, which at the time was a peer-to-peer network only, with no exchanges or wallet providers. The transaction marked a significant turning point in the growth of the Bitcoin network and demonstrated that Bitcoin could be successfully moved. Hal Finney contributed significantly to the growth of Bitcoin and the larger cryptocurrency business. He was one of the first Bitcoin adopters. He died in August 2014 from ALS-related complications.

WHAT IS A BLOCKCHAIN?

A blockchain is a distributed, open ledger that keeps track of every transaction that occurs within a network. The blockchain in the context of Bitcoin acts as a visible and unchangeable record of each Bitcoin transaction. The blockchain is created by grouping every transaction into blocks and connecting them in chronological order. A network of users, referred to as nodes or miners, maintains the blockchain by validating and verifying transactions. This distributed structure guarantees that no one entity has control over or access to the transaction history. Cryptographic techniques are used to maintain the blockchain's security, which makes it extremely resilient to fraud and tampering. The blockchain concept has been used in many different businesses and use cases, providing chances for openness, effectiveness, and trust in diverse fields.

HOW IS BITCOIN DIFFERENT FROM TRADITIONAL CURRENCIES?

Bitcoin is unique from conventional currencies in a number of ways. First of all, unlike centralized entities like governments or central banks, Bitcoin is decentralized and runs independently of them. It is not connected to any one nation or legal system. Second, only a certain number of Bitcoins are issued each day. Traditional currencies are susceptible to inflation as a result of an expanding money supply, while Bitcoin has a maximum supply of 21 million coins, which fosters scarcity and may prevent inflation. Thirdly, the security of Bitcoin transactions is provided by cryptographic technology. Last but not least, Bitcoin has the capability of pseudonymity, enabling users to conduct transactions without disclosing their real names. Many jurisdictions still have laws and regulations that apply to Bitcoin.

CAN I ACQUIRE BITCOIN, AND IF SO, HOW?

Purchasing Bitcoin from cryptocurrency exchanges is one option. Through these platforms, you can buy Bitcoin with conventional fiat money like the US dollar, euro, peso, rand, etc. You will need to register for an account, go through the required verification steps, and add money to the exchange. You can place an order to purchase Bitcoin once your account has been financed. Bitcoin can also be obtained by mining. Large-scale operations and specialized mining equipment, however, are used for mining, although it's possible to do small-scale operations. As an alternative, you can get paid in Bitcoin for products or services. Nowadays, a large number of companies and people accept Bitcoin as payment. Through tools like RaiseSats, you can crowdfund Bitcoin or receive

Bitcoin as a gift or donation.

WHAT IS A BITCOIN WALLET, AND HOW DO I SET ONE UP?

A computer instrument called a Bitcoin wallet is used to handle, store, and carry out transactions with Bitcoin. It has two crucial parts: a private key that gives you access to your Bitcoin and a public key that acts as your wallet address. Different kinds of Bitcoin wallets exist. While hardware wallets are actual physical devices that offer higher protection, software wallets may be loaded on your computer or mobile device. Web wallets are online services that enable you use a web browser to access your Bitcoin. Regardless of the choice, it is essential to safeguard your private key and backup your wallet to reduce the chance of losing your Bitcoin.

IS BITCOIN ANONYMOUS?

Transactions made using Bitcoin are pseudonymous, which means they are not directly connected to a specific individual's identity. Transactions are instead linked to Bitcoin addresses, which are character sequences produced at random. Even while Bitcoin has some privacy features, it is not entirely anonymous. Transactions can be tracked and studied because the blockchain is open-source. Your transactions may be tracked if someone can connect your Bitcoin address to your identity. Additionally, when using exchanges that have Know Your Customer (KYC) regulations, using Bitcoin can leave digital traces. Some people use methods to increase their privacy, such as using numerous addresses, using mixing services, or VPNs with Tor. The use of Bitcoin in illicit activity is still subject to investigation and prosecution. It is worth noting that more crime is done with fiat currency or government issued currency than is done with Bitcoin or cryptocurrency in general.

HOW CAN I ENSURE THE SECURITY OF MY BITCOIN?

The most important thing is to not trust but to verify. Utilize a reputable Bitcoin wallet from an established vendor while keeping this in mind. Choose a wallet after doing some research that has a solid reputation for security and good user ratings. It is crucial that your wallet and any related accounts support two-factor authentication (2FA). 2FA increases security by requiring a second verification step, typically through a different device or app. Additionally, avoid using easily guessed information and make sure your Bitcoin wallet password is secure and unique. Use a hardware wallet, which offers offline storage and other security precautions. Always use caution when exchanging your private key or conducting transactions. Be cautious of phishing scams and dangerous websites.

WHAT IS A PUBLIC KEY HASH?

An abbreviated version of a Bitcoin public key is known as a hash. It is created via hashing, a cryptographic technique, from the public key. Bitcoin addresses, which are often displayed as a collection of letters and numbers to make them easier for users to handle and share, are created using the public key hash. The public key's security and integrity are maintained while allowing for more effective storage and quicker transaction processing.

WHAT IS A PRIVATE KEY, AND WHY IS IT IMPORTANT?

A randomly generated string of characters known as a private key is mathematically connected to the public key of your Bitcoin wallet. It is an essential component that lets you to sign transactions and gain access to your Bitcoin. Given that anyone with access to your private key can manage your Bitcoin, private keys are designed to be kept hidden and secure. It's critical to maintain your private key safely because misplacing it could result in irreversible loss of access to your Bitcoin. Writing it down on paper and storing it safely, using a hardware wallet that securely keeps your private keys offline, or using a password manager are some typical techniques. Your Bitcoin is vulnerable to theft if someone gets their hands on your private key.

WHAT IS A BITCOIN TIMESTAMP?

A Bitcoin transaction that has a specified time and date is referred to as having a timestamp. A timestamp that identifies the precise minute a transaction was produced and broadcast to the network is included in every transaction in the Bitcoin network. The timestamp aids in determining the chronological order of transactions within the blockchain along with other transaction information. This function guarantees that transactions are executed in a particular order, giving a trustworthy account of what happened on the Bitcoin network.

WHAT IS A SEED PHRASE?

A Bitcoin mnemonic seed is a set of words that acts as a backup for a Bitcoin wallet. It is also referred to as a recovery seed or seed phrase. A set of 12 or 24 randomly generated words are often given to consumers when they create a Bitcoin wallet. If the original wallet file or device is lost, broken, or otherwise not accessible, these phrases, when typed in the proper order, can be used to regain access to the wallet. The mnemonic seed offers a simple and consistent method for backing up and recovering Bitcoin wallets, hence boosting user security and guaranteeing wallet accessibility.

CAN I LOSE MY BITCOIN?

If you don't take the necessary safeguards, you could lose your Bitcoin. You may lose access to your Bitcoin if you misplace your private key, your wallet is compromised, or both. There is no way to retrieve your private key if you lose it, and your Bitcoin will be permanently lost if you don't have a backup. You might not be able to access your Bitcoin if you lose the physical storage device or forget the password to your wallet. Keep regular backups of your wallet, and save your private key or seed phrase safely in several secure places. To reduce the chance of loss, make sure you are using trusted wallet software and adhere to basic security standards. Keep in mind that Bitcoin transactions cannot be undone.

HOW ARE BITCOIN TRANSACTIONS VERIFIED?

Mining is the method through which Bitcoin transactions are validated and added to the network. A transaction is broadcast to the network of involved nodes when it is started. Through a procedure known as proof-of-work, miners compile these transactions and compete to solve challenging mathematical riddles. The computational effort and resources needed to solve this puzzle are considerable. When a miner completes the problem, they broadcast the answer to the network so that other nodes can validate and confirm it. The miner adds the block containing the transaction to the blockchain after the network has confirmed the solution. This procedure verifies the legitimacy of transactions and avoids double-spending. By adding a new block to the blockchain, it strengthens network security and verifies the inclusion of the transaction in the ledger.

WHAT IS MINING
IN THE CONTEXT
OF BITCOIN?

The process of validating and adding transactions to the blockchain as well as producing new Bitcoins are collectively referred to as "mining" in the context of Bitcoin. Participants who participate in mining use powerful computers to tackle challenging mathematical riddles. Miners secure the network and validate the legitimacy of transactions by resolving these riddles. The freshly created Bitcoin and any associated transaction fees are given to the miner who successfully completes the puzzle. Due to its resource-intensive nature and high computational complexity, mining calls for specialized hardware known as ASICs (Application-Specific Integrated Circuits). It is essential for preserving the safety and reliability of the Bitcoin network, ensuring that all transactions are legal that the blockchain is impervious to manipulation.

HOW DOES BITCOIN MINING WORK?

Proof-of-work is a technique used in Bitcoin mining. A computationally challenging mathematical challenge is solved by miners in competition. It is intended to take a lot of processing power to do this, which costs money. In an effort to locate the answer, miners employ specialized hardware and strong computers that can process many calculations per second. When a miner discovers a solution, they broadcast it to the network so that other miners can validate and verify it. Once the transactions have been verified, the successful miner adds a new block of transactions to the blockchain, receives a reward in the form of freshly created Bitcoin, and gains transaction fees from the transactions that were added. To maintain a constant block generation rate, mining difficulty is adjusted. Mining becomes harder as the network expands, needing more power and energy. It is worth noting that many have cried out about the energy consumption of Bitcoin but many Bitcoin mining efforts are shifting to mining with green energy such as solar, wind, hydro, or even volcanic energy.

WHAT IS A BITCOIN EXCHANGE, AND HOW CAN I BUY BITCOIN ON ONE?

A Bitcoin exchange is a website where you can purchase, sell, and exchange Bitcoin for other cryptocurrencies or conventional fiat money. These marketplaces serve as brokers, connecting buyers and sellers and streamlining the transaction procedure. On most Bitcoin exchanges, in order to purchase Bitcoin, you must first register for an account, go through the required verification steps, then fund your account. You can either establish a specific price at which you want to buy or submit a buy order for Bitcoin at the current market price once your account has been financed. When the transaction is finished, the Bitcoin will be credited to your exchange wallet. The exchange will carry out the deal on your behalf. Choosing a trustworthy exchange with high security is crucial but above all, holding your own Bitcoin is key because if an exchange goes down for maintenance, whatever is left on it is inaccessible. In the early 2010s, the largest cryptocurrency exchange in the world was Mt. Gox, which had its headquarters in Japan. Jed McCaleb started it in 2010, and in 2011 Mark Karpeles acquired it. Over the years, Mt. Gox experienced numerous security lapses and cyberattacks, which led to the loss of hundreds of millions of dollars in client assets. Many users were unable to access their Bitcoin after Mt. Gox abruptly stopped operating and filed for bankruptcy in February 2014. According to the bankruptcy procedures, Mt. Gox had lost almost 850,000 Bitcoins, which is a staggering sum of money. The demise of Mt. Gox was a pivotal moment in the development of Bitcoin, emphasizing the significance of self-custody.

CAN I BUY FRACTIONS OF A BITCOIN?

Bitcoin can be split into smaller units that can be purchased. The smallest Bitcoin unit is referred to as a "satoshi," after the anonymous Bitcoin inventor Satoshi Nakamoto. 100 million satoshis make up one Bitcoin. Due to its divisibility, Bitcoin provides for flexible transaction sizes that can handle both small and large transactions. You can indicate how much Bitcoin you wish to buy when doing so, whether you want to buy a full Bitcoin or only a portion of one. Exchanges and wallets normally show the price of Bitcoin in terms of one Bitcoin, but you can enter any quantity, even fractions, that you want to purchase. You can purchase fractions of Bitcoin as tiny as 0.001 or as large as 0.1 or 0.5.

ARE THERE ANY FEES ASSOCIATED WITH BITCOIN TRANSACTIONS?

There are costs involved with transactions using Bitcoin. The miners that include and verify your transaction in a block receive a transaction fee each time you send Bitcoin from one address to another. The transaction fee provides miners with a reward for giving your transaction priority and including it right away. The charge can change depending on the transaction's urgency, the size of the transaction, and other variables including network congestion. While lower-fee transactions may face delays, higher-fee transactions often execute transactions more quickly. The cost of the transaction is reduced by the fee, which is paid in addition to the money you transmit. Depending on your choices and needs, wallets will either estimate the charge or let you manually select it.

CAN I SEND BITCOIN TO ANYONE IN THE WORLD?

With an internet connection and a Bitcoin wallet, you can send Bitcoin to anyone in the world. One benefit of Bitcoin is its borderlessness, which enables quick and international transactions without the use of middlemen. You need the recipient's Bitcoin address, which is a distinctive identifier for their Bitcoin wallet, in order to send Bitcoin to them. You can start a transaction from your wallet and specify the amount of Bitcoin you want to send after you know the recipient's Bitcoin address. The Bitcoin network will publish the transaction, and miners will then confirm it. The funds will appear in the recipient's wallet as soon as the transaction is approved. Before beginning a transaction, make sure the recipient's address is correct to prevent any mistakes. It's worth noting that new ways are emerging where you can

leverage mesh networks to send offline Bitcoin transactions.

HOW LONG DOES IT TAKE TO SEND BITCOIN?

As a result of high demand and a large number of pending transactions in the Bitcoin network, network congestion is a significant factor. Transactions could incur delays and require more time to confirm during periods of high network activity. Transactions with greater transaction fees are often completed more quickly because miners give priority to those transactions depending on the transaction cost. Approximately every 10 minutes, Bitcoin strives to generate a new block, but actual confirmation times for transactions can vary. If you charge a fair transaction fee, your transaction will be more likely to be accepted into the following block and processed more quickly. You can get an indication of how long your transaction would take from some wallets or exchanges that estimate the time it will take for confirmation of a transaction depending on the state of the network.

WHAT IS A BITCOIN ADDRESS, AND CAN I CHANGE IT?

A destination or recipient for Bitcoin transactions is represented by a unique identifier known as a Bitcoin address. It is a collection of alphabetic characters produced by cryptography methods. The public key linked to a Bitcoin wallet is used to generate Bitcoin addresses, which are then used to receive Bitcoin. Each Bitcoin address is distinct and can be used to send and receive money from other users. A private key, which is required to access and manage the Bitcoin linked to an address, is not part of a Bitcoin address. A Bitcoin address cannot be directly changed, however new Bitcoin addresses can be created. To improve security and anonymity, many wallets create a new address automatically for each transaction. It's advised to send the recipient a fresh

Bitcoin address.

IS BITCOIN LEGAL?

Each country has its own laws governing. Bitcoin may be subject to limitations or outright bans in some areas, while others have full legalization and regulation of it. Due to its absence from established financial institutions and regulations, Bitcoin typically functions in a legal limbo. While some nations have embraced Bitcoin and accepted it as a valid form of money or commodity, others have adopted a more cautious stance. It's crucial to learn about and comprehend your jurisdiction's legal and regulatory framework with reference to Bitcoin. To make sure you are using Bitcoin legally and in compliance with local regulations, adhere to all tax and reporting duties.

CAN BITCOIN BE REGULATED?

Around the world, different countries have adopted different approaches to Bitcoin regulation. Governments have imposed rules on the organizations and activities that surround Bitcoin, such as exchanges and financial institutions, despite the fact that Bitcoin itself is decentralized and not governed by any one centralized body. The purpose of regulation is to stop money laundering, illegal activity, and to safeguard consumers. Some nations have enacted licensing and registration specifications for companies doing business in the Bitcoin industry. Know-your-customer (KYC) and anti-money laundering (AML) processes, reporting requirements, and taxation regulations are examples of regulatory measures. It's important to remember that regulations are still changing and might vary greatly different nations. For compliance and to guarantee the legal usage of Bitcoin, it's essential to stay up to date on

legal and regulatory developments in your jurisdiction.

CAN I USE BITCOIN FOR ILLEGAL ACTIVITIES?

Bitcoin is a neutral technology and can be used for both legitimate and illegitimate activities. While Bitcoin provides some anonymity, the public blockchain makes transactions traceable, therefore it is not entirely anonymous. It is a myth that only unlawful activity is carried out using Bitcoin. The bulk of Bitcoin users utilize it for legitimate business dealings and regard it as a digital asset or method of payment. Bitcoin can, however, be misused by anyone engaged in unlawful activity, just like any other financial system. It's crucial to remember that law enforcement authorities have the tools and methods necessary to track down and look into shady Bitcoin transactions. It is against the law and may have legal repercussions to use Bitcoin for illicit activities.

CAN I USE BITCOIN FOR EVERYDAY PURCHASES?

The variety of companies and shops that accept Bitcoin has grown over time. Numerous products and services, including online purchases, reservations for trips, gift cards, and even actual goods from some merchants, can be purchased using Bitcoin. Merchants can take Bitcoin by converting it into local currency or by accepting Bitcoin payments directly thanks to payment processors and services. Using Bitcoin for regular transactions is also made simpler by mobile wallets and payment applications. Bitcoin is still not as commonly accepted as conventional payment methods, and acceptance may vary by location, it's crucial to remember that. Make sure the retailer or service provider accepts Bitcoin as payment before making a purchase.

HOW CAN I SELL MY BITCOIN?

One of the best ways to sell your Bitcoin is by using a Bitcoin exchange or a service that allows withdrawals in fiat money, peer-to-peer trading platforms, or two-way Bitcoin ATMs. You can issue a sell order indicating how much you want to sell for and at what price if you own Bitcoin on an exchange. The exchange will handle the transaction whenever a buyer matches your sell order, and the Bitcoin will be sent from your account to the buyer's account. You can either withdraw the proceeds of the sale to your bank account or utilize them to buy other cryptocurrencies. As an alternative, you can use peer-to-peer trading systems to sell Bitcoin directly to other people. You can negotiate the conditions of a deal using these online marketplaces that link buyers and sellers.

HOW DOES THE PRICE OF BITCOIN CHANGE?

The dynamics of market supply and demand determine the price of Bitcoin. It is fluctuating and capable of quick changes. Market mood, investor demand, macroeconomic conditions, regulatory changes, media coverage, and technological improvements are some of the elements that affect the price of Bitcoin. Changes in price can also be attributed to Bitcoin's finite quantity and its capacity to serve as a store of wealth. The trading activity on cryptocurrency exchanges, where buyers and sellers join together to determine the market price, is what essentially determines the price of Bitcoin. Large buy or sell orders, as well as market orders from several participants, can affect the price. It's vital to keep in mind that the cryptocurrency market is extremely volatile and that the price of Bitcoin can change dramatically in a short amount of time.

WHAT IS BITCOIN MINING DIFFICULTY?

The term "Bitcoin mining difficulty" refers to a measurement of how challenging it is to solve the mathematical puzzles necessary to add new blocks to the blockchain. To maintain a constant block creation rate of roughly one block every 10 minutes, the Bitcoin network increases the mining difficulty every two weeks on average. Based on the network's overall computational capacity, the mining difficulty is changed. The difficulty grows as more miners join the network and add more processing power. The difficulty goes down as miners leave the network. This guarantees that new blocks are added to the blockchain at a roughly consistent rate, independent of the network's overall hash rate. It is difficult to manipulate the blockchain because of the mining difficulty, which upholds the reliability and security of Bitcoin.

IS BITCOIN A GOOD INVESTMENT?

Bitcoin is a brand-new asset with ascending value that has recently come into the public view. Its price has increased significantly over time, indicating that it is an asset that merits attention. Having said that, Bitcoin's price can change drastically in a short amount of time and can be quite volatile. There is a risk associated with investing in Bitcoin, including the possibility of capital loss. Before making an investment decision, it is recommended to do extensive research, understand the dangers involved, and take into account getting competent financial guidance. Additionally, risk management and diversification are crucial components of an investment strategy. When thinking about investing in Bitcoin, it's crucial to set a budget that you can afford to lose and to think long term. Dollar cost averaging is one of the best methods for all, no matter how much money you have, as it involves taking a budgeted amount of capital and deploying it over a set time horizon. An example of this is to buy $10 of Bitcoin for 100 days, no matter what the price. With this, you are able to circumnavigate market volatility and attempting to "time the market" by accumulating based on time *in* the market.

CAN I LOSE MONEY INVESTING IN BITCOIN?

Yes, there is a chance to lose money when investing in Bitcoin. Bitcoin's price has the potential to be extremely erratic, and its value may change significantly. The price of Bitcoin has the capacity to rise, but it also has the potential to fall, occasionally sharply. Market risk, liquidity risk, regulatory risk, and technology risk are just a few of the risks associated with investing in Bitcoin. Market mood, the state of the economy, changes in regulations, and technological improvements are just a few examples of the variables that might affect Bitcoin's price. Before investing in Bitcoin, it's crucial to thoroughly weigh these dangers, assess your risk appetite, and analyze your financial status. To spread the risk, diversify your investing portfolio and only invest money you can afford to lose.

WHAT IS THE DIFFERENCE BETWEEN BITCOIN AND ALTCOINS?

The first cryptocurrency and still the most popular and frequently used is Bitcoin. It runs on its own blockchain and is frequently referred to as digital gold or a store of value. On the other side, cryptocurrencies other than Bitcoin are referred to as altcoins. In comparison to Bitcoin, altcoins may have various features, use cases, and underlying technology. They might provide enhancements or alternative functionality outside the realm of Bitcoin. Among the many different altcoins are Ethereum, Litecoin, Ripple, Stacks, and many more. Some alternative currencies (altcoins) are designed to meet particular industry demands or provide DApp platforms. While Bitcoin continues to hold a sizable market share and maintain its dominance, altcoins provide a wide range of cryptocurrency options, each with its own advantages and possible uses. Bitcoin served as a launching point for many altcoin users.

WHAT IS THE MAXIMUM SUPPLY OF BITCOIN?

21 million coins are the maximum number of Bitcoins in circulation. The original Bitcoin whitepaper describes this fixed supply. Through the mining process, new Bitcoins are gradually added to the supply. For each block they add to the blockchain, miners are given a set amount of Bitcoins as payment. 50 Bitcoins per block were the initial values for the block reward. The block reward is intended to diminish over time through a process known as the Bitcoin halving. The block prize is divided in half around every four years. This halving occurrence continues until there are 21 million Bitcoins available overall, which is anticipated to occur in the year 2140. The asset's scarcity is increased by Bitcoin's constrained supply.

WHAT HAPPENS WHEN ALL THE BITCOINS ARE MINED?

The block reward for miners will be zero after all of the Bitcoins have been mined, which is anticipated to occur around the year 2140. Even though miners continue to play a crucial role in the network by validating and verifying transactions, they will no longer get freshly created Bitcoins as payment for adding blocks to the blockchain. Transaction fees will serve as their primary source of motivation for their mining operations. People using Bitcoin pay transaction fees as a compensation for miners, encouraging them to keep running their operations and protecting the network. As the Bitcoin network adapts to this new economic model, the switch from block rewards to transaction fees is anticipated to occur gradually.

CAN BITCOIN BE COUNTERFEITED?

Bitcoin is nearly impossible to counterfeit because to its design and cryptographic concepts. The blockchain is updated and confirmed for every Bitcoin transaction through a process involving intricate mathematical computations. The blockchain's transparency and immutability guarantee the accuracy of the transaction history. The proof-of-work consensus mechanism used by Bitcoin ensure that adding new blocks to the blockchain is computationally demanding and needs a lot of computing power. Due to the decentralized nature of the network and the mining process, it is extremely unlikely that a single person, organization, or group will counterfeit or manipulate the Bitcoin supply. The risk of fake Bitcoin is almost nonexistent as long as the majority of miners, who are economically incentivized, honest, and follow the decentralized standard.

CAN BITCOIN BE HACKED?

Individual Bitcoin wallets and exchanges can be subject to hacking attempts. Thanks to its strong cryptographic foundation and decentralized network, Bitcoin has not been compromised since its inception. However, exchanges and wallets have experienced security lapses, which resulted in Bitcoin being stolen. Hackers use a variety of techniques, such as phishing, malware, social engineering, and exploiting holes in infrastructure or software. It is essential to utilize trustworthy and secure wallets and exchanges, enable two-factor authentication, routinely update software, and adhere to best security practices including using strong passwords and being wary of dubious links or emails in order to reduce the danger of hacking. Remember: don't trust but verify.

WHAT IS A BITCOIN FORK, AND HOW MANY FORKS OF BITCOIN ARE THERE?

When the protocol or rules governing the Bitcoin network are significantly altered, the existing blockchain diverges, and this is referred to as a "Bitcoin fork." When there are disagreements among members of the Bitcoin community over updates, modifications, or the network's overall direction, forks might result. Both soft and hard forks fall into this broad category. A hard fork introduces modifications to the protocol that are not backwards compatible and creates a new blockchain, as opposed to a soft fork that introduces changes that are. By altering the original Bitcoin code, hard forks can result in the formation of brand-new coins. Bitcoin forks include, for instance,

Bitcoin Cash (BCH), Bitcoin SV (BSV), and Bitcoin Gold (BTG).

Confusion frequently occurred from these forks, which were

not generally accepted by many Bitcoin purists.

WHAT IS A BITCOIN HALVING?

The block reward provided to miners is halved as part of a Bitcoin event that happens about every four years. It is built into the Bitcoin protocol as a mechanism to regulate the creation of new Bitcoins and eventually slow down the rate at which they enter the market. When Bitcoin first came online in 2009, the block reward was set at 50 Bitcoins. The prize was reduced by the first halving to 25 Bitcoins in 2012. The second halving took place in 2016, bringing the payout down to 12.5 Bitcoins. The block reward was reduced by the most recent halving to 6.25 Bitcoins in May 2020. Bitcoin's monetary policy includes periodic halving events, which add to its scarcity.

CAN I EARN INTEREST ON MY BITCOIN HOLDINGS?

You can use a variety of strategies to earn interest on your Bitcoin holdings. One well-liked approach involves Bitcoin lending marketplaces, which link borrowers and lenders and enable Bitcoin owners to lend their coins to borrowers in exchange for interest payments. The platform, the length of the loan, and the creditworthiness of the borrower are just a few examples of the variables that might affect interest rates. By putting your Bitcoin into these platforms, you can use loan pools or liquidity mining programs to generate income on your investments. Participating in lending and DeFi (Decentralized Finance) platforms entails risks, such as the possibility of losing money due to platform flaws or defaults. Also, Stacks, a Layer 2 blockchain that has a unique relationship with Bitcoin, offers Bitcoin yield based on it's proof-of-transfer consensus mechanism. Assessing tax exposure and doing due diligence are of the utmost significance. For many, simply possessing Bitcoin is sufficient but more options are available.

WHAT IS A BITCOIN ETF?

Without actually holding or owning Bitcoin, investors can still acquire exposure to the digital currency by purchasing a Bitcoin exchange-traded fund (ETF). A Bitcoin ETF is made to track Bitcoin's price fluctuations and give investors a method to purchase Bitcoin through conventional investing avenues like stock exchanges. According to the ETF's structure, investors can purchase shares or units of the fund, which correspond to a portion of the fund's underlying Bitcoin holdings. This gives investors access to Bitcoin's price performance in a convenient and controlled way without having to deal with the hassles of actually owning and safeguarding the cryptocurrency. Because they are traded on conventional stock exchanges and are subject to regulatory approval, Bitcoin ETFs are available to a wide range of investors.

CAN I PURCHASE BITCOIN IN MY RETIREMENT ACCOUNT?

Some retirement plans, such a Solo 401(k) or an IRA that is self-directed, allow for the purchase of Bitcoin (k). These retirement account options give investors greater discretion over their investing decisions and support a wider variety of assets, such as Bitcoin and other cryptocurrencies. You would normally need to set up a self-directed account with a competent custodian that supports cryptocurrency investments in order to purchase Bitcoin in a retirement account. In order to comply with the unique rules and specifications for retirement accounts, the custodian will facilitate the acquisition and storage of Bitcoin within the account. Cryptocurrency investments are not allowed in all retirement accounts. To learn about the ramifications, dangers, and taxes related to investing in Bitcoin, speak with a financial counselor or tax expert.

CAN I USE BITCOIN AS COLLATERAL FOR A LOAN?

Individuals can borrow money by using their Bitcoin as collateral for Bitcoin-backed loans, also known as crypto-backed loans or Bitcoin collateralized loans. A specified quantity of Bitcoin must normally be deposited into the safe wallet of a lending platform before the transaction can proceed. Depending on the value of the collateral, borrowers may then get a loan in the form of traditional money or other currencies. Depending on the lending platform, different loan terms apply, including different interest rates, loan-to-value ratios, and repayment plans. Without having to liquidate your Bitcoin holdings, using Bitcoin as collateral for a loan can provide liquidity, allowing you to profit from any future price growth. Before agreeing to any lending agreements, do your homework and make sure you are aware of all the conditions, dangers, and fees related to loans

backed by Bitcoin.

DO PEOPLE SCAM FOR BITCOIN?

Scammers try to steal value from individuals or systems, and at the moment, Bitcoin is the biggest digital asset available to users worldwide. Knowing what kind of frauds are frequently used may help you be vigilant. Fraudulent emails, websites, or messages that imitate reliable platforms or services are used in phishing schemes. Users are tricked into divulging their private keys, passwords, or other sensitive information by con artists. Before entering any personal or financial information, always make sure a website and communication channel are legitimate. Ponzi schemes use money from new investors to pay money to earlier investors, promising people huge returns on their investments. When the flow of new investor monies stops, these scams eventually fall apart. Investment opportunities with unrealistic or guaranteed profits should be avoided.

The purpose of phony cryptocurrency exchanges and wallet services created by con artists is to steal your Bitcoin or personal data while making them appear authentic. Use trusted wallets and exchanges, and make sure to check their legitimacy before using either. Be wary of people or businesses that advertise assured returns or privileged investing opportunities. If an investment opportunity seems too good to be true, do your research, check out the credentials, and be skeptical. Theft of Bitcoin private keys or login credentials from infected devices is possible via malicious software or key loggers. While downloading or dealing with unidentified files or links, users should use current antivirus software to keep their devices secure. Social engineering techniques may be used by hackers to trick people into giving them access to their Bitcoin wallets. Be wary of incoming correspondence or inquiries requesting private information. Hackers may target centralized exchanges and online wallets, resulting in the theft of Bitcoin assets. Use trusted exchanges, wallets with strong security features like two-factor authentication, and cold storage, as well as two-factor authentication. The phrase that pays is "Don't trust, verify."

CAN BITCOIN BE USED FOR EVERYDAY PURCHASES?

Bitcoin acceptance is currently less than that of other established payment methods, but it can be utilized for regular purchases. The number of companies and merchants that now accept Bitcoin as payment has increased over time. They include not only online retailers but also physical shops, eateries, and even certain service providers. Each retailer that accepts Bitcoin payments will let you use it to pay for products, services, or even gift cards. A Bitcoin wallet is typically require. To complete the transaction, you can give the merchant your wallet's generated Bitcoin address, which is specific to each transaction. The transaction is subsequently published to the Bitcoin network and included to a block for confirmation. Most people who use Bitcoin daily leverage the Lightning Network for fast transactions.

Be mindful that there are regional and merchant differences in the acceptance of Bitcoin.

CAN I USE BITCOIN FOR INTERNATIONAL TRANSACTIONS?

Bitcoin may be used for international transactions, and its universal accessibility is one of its benefits. Bitcoin transactions can be carried out between persons in various nations without the use of middlemen like banks or payment processors. Direct peer-to-peer transactions are made possible by Bitcoin's decentralized structure, allowing for quick and limitless value transfers. No matter where the recipient is physically located, a Bitcoin transaction may be started by only knowing their Bitcoin address. When utilizing Bitcoin for overseas transactions, it's crucial to keep things like transaction costs, currency exchange rates, and regulatory regulations in mind. When making cross-border Bitcoin transactions, it's also crucial to adhere to the laws and tax regulations of both your country and the country of the

recipient.

CAN I USE BITCOIN FOR REMITTANCES?

Remittances, or the transfer of money from one country to another, can be done using Bitcoin. Usually, people utilize remittances to send money to their friends and family back home. Individuals find Bitcoin to be a desirable choice due to its lack of geographical restrictions and lower transaction costs than more established remittance methods. Bitcoin makes international money transfers quicker and more affordable, especially in places with weak traditional banking infrastructure. However, it's crucial to take into account the variables that may affect the price and effectiveness of Bitcoin remittances, including exchange rates, transaction costs, and the availability of Bitcoin in the country of the recipient. The simplicity of sending and receiving Bitcoin via remittances is impacted by regulatory regulations and the accessibility of Bitcoin infrastructure, which differ between nations.

CAN I USE BITCOIN TO PAY MY BILLS?

In some circumstances, you may be able to pay your expenses with Bitcoin. Some service providers and platforms now accept Bitcoin as a means of payment, albeit it is still less common than traditional payment methods for paying bills. Online shops, utility providers, and even some governments can be examples of them. In order to pay your bills with Bitcoin, you normally need to find a platform or service that accepts the currency and adhere to their detailed instructions. This can be done by manually entering the recipient's Bitcoin address and the desired payment amount or by scanning a QR code. Payments made using Bitcoin may be subject to additional fees or exchange rate considerations. Check the specific service provider's terms and conditions, costs, and available methods of payment.

IS BITCOIN TAXED?

Bitcoin is subject to taxes in various juridictions. Instead of being treated as legal money, tax authorities view Bitcoin and other cryptocurrencies as assets or property. The way that Bitcoin is taxed depends on the nation and its own tax regulations. Bitcoin sales or exchanges may generate profits that are taxed as capital gains. The capital gain is normally taxable if you hold Bitcoin as an investment and sell it for more money than you paid for it initially. The holding period and the local tax regulations are just two examples of the variables that affect the tax rate. The value of any Bitcoin you get is typically taxable as income if it is used to purchase goods or services, pay for a wage, or is included in self-employment income. The fair market value of the Bitcoin at the moment of receipt determines the tax liability. People must disclose their Bitcoin holdings and transactions to tax authorities in various countries, either on their tax returns or through additional reporting channels. Penalties and legal repercussions may arise if these reporting requirements are not followed. The value of newly mined Bitcoin may be taxable as income if you mine Bitcoin. Furthermore, any benefits derived from taking part in mining pools or staking activities can potentially be subject to taxation. Keeping correct records and reporting Bitcoin transactions are essential for meeting your tax requirements and avoiding fines.

WHAT IS A BITCOIN WALLET BACKUP, AND WHY IS IT IMPORTANT?

An encrypted backup of your wallet's private keys or seed phrase is referred to as a backup of your Bitcoin wallet and is kept there in case you need to recover or restore it. For managing and controlling your Bitcoin, you must have the private keys. The permanent loss of your Bitcoin holdings may occur if you misplace your private keys. To protect your money from risks including device failure, unintentional deletion, and theft, it is essential to have a backup wallet. In the event of any unanticipated events, it enables you to recover your wallet and get back access to your Bitcoin. It's also advised that you frequently update your backup as you create new addresses or private keys to make sure you can still access your Bitcoin in the future.

CAN I RECOVER MY BITCOIN IF I LOSE MY WALLET?

Recovering your Bitcoin if you lose your wallet depends on whether you have a backup of your wallet's private keys or seed phrase. If you have a backup, you can restore your wallet and regain access to your Bitcoin. Most Bitcoin wallets provide an option to restore a wallet using a backup by importing the private keys or entering the seed phrase. By following the wallet's instructions, you can recover your wallet and access your Bitcoin again. It's important to emphasize that your private keys or seed phrase are critical to wallet recovery. If you do not have a backup and lose access to your wallet without any means to recover your private keys, it can result in permanent loss of your Bitcoin. Therefore, it is crucial to create a backup and store it securely in multiple locations to prevent the risk of losing access to your Bitcoin.

HOW CAN I KEEP MY BITCOIN SAFE FROM HACKERS AND THEFT?

To preserve your digital assets, you must protect your Bitcoin against theft and hackers. Use a reliable Bitcoin wallet: Do your research and select a wallet from a respected vendor with a solid reputation for security. Turn on two-factor authentication (2FA): To add an additional layer of security, turn on 2FA for your wallet and any related accounts. Create a strong password for your wallet and steer clear of utilizing information that is simple to decipher. Maintain your software: Update your wallet software frequently to ensure you have the most recent security patches and bug fixes. Take into account utilizing a hardware wallet: These wallets offer offline storage and additional security measures, making them less vulnerable to internet attacks. Phishing attempts should be avoided: Avoid clicking on dubious links, phishing emails, and rogue websites that can jeopardize your Bitcoin security. Always be sure the sources are reliable, and never divulge critical information. Don't lose your wallet: Make copies of the private keys or seed phrase for your wallet and store them safely in several places. This makes sure that if your wallet is lost or damaged, you may still recover your Bitcoin. Keep up with the most recent security procedures and emerging dangers in the cryptocurrency industry. To remain ahead of possible hackers, keep current with typical attack methods and security precautions.

CAN I USE THE SAME PASSWORD FOR ALL MY BITCOIN-RELATED ACCOUNTS?

It's not advised to use the same password for all of your Bitcoin-related accounts. Although it might be practical, a security breach is much more likely as a result. Your other accounts, including your Bitcoin wallet, could all be accessed if a hacker manages to get into one of your accounts and figure out your password. For each of your accounts, including your Bitcoin wallet, exchange accounts, and any other platforms connected to Bitcoin, it's crucial to have a different, secure password. Using a mix of capital and lowercase letters, numbers, and special characters will help you create strong passwords. Consider utilizing a password manager as well to create strong passwords for all of your accounts and store them securely. You increase overall

Bitcoin security and lessen the effects of any future breaches.

CAN I RECOVER A FORGOTTEN BITCOIN WALLET PASSWORD?

Given that the majority of wallet providers prioritize security and do not have access to your password, recovering a forgotten Bitcoin wallet password can be difficult. Wallets do not save your password or have the capability to reset it because they are made to prevent unauthorized access. The only option if you forget your wallet password may be to employ a wallet recovery mechanism, like a backup file or a seed phrase. You can reset your password or restore your wallet to a different device using these recovery methods. There could be no way to retrieve your Bitcoin if you lose your recovery options and forget your password. Remember your password and keep recovery options securely to prevent losing access to your Bitcoin.

WHAT IS THE DIFFERENCE BETWEEN A HOT WALLET AND A COLD WALLET?

An internet-connected hot wallet, sometimes referred to as an online wallet, enables easy and rapid access to your Bitcoin. Hot wallets are primarily software-based wallets that may be accessible by web-based interfaces, desktop or mobile applications, or both. They are excellent for frequent access to your Bitcoin and offer simplicity of usage. Hot wallets may, however, be more susceptible to online threats and hacking attempts because they are connected to the internet. When utilizing a hot wallet, use strong security precautions, such as setting two-factor authentication. A cold wallet, sometimes referred to as an offline wallet or a hardware wallet, is the most secure choice because it is not online. Your Bitcoin is normally safely stored offline on a physical medium or piece of hardware. Because they are immune to online threats, cold wallets provide increased protection. They are perfect for holding large sums of Bitcoin or for long-term storage. Cold wallets shield your Bitcoin from hacking attempts and unwanted access by keeping it offline. However, in order to access Bitcoin from a cold wallet, you must either move money to a hot wallet beforehand or use the cold wallet in conjunction with appropriate software. Your own demands and tastes will determine whether you choose a hot wallet or a cold wallet. While cold wallets place a higher priority on security and protection from online threats, hot wallets provide accessibility and ease. Many people opt to utilize both in conjunction.

WHAT IS A PAPER WALLET?

A paper wallet is a tangible paper document that holds the data you need to access and manage your Bitcoin. Your Bitcoin address and the related private key or seed phrase are often included. The private key is protected from potential online threats by the offline generation of paper wallets. Because they are kept offline, they are regarded as a type of cold storage, reducing the possibility of hacking or unwanted access. Paper wallets are frequently used because they offer a high level of security for long-term storage of Bitcoin. When you need to access your Bitcoin, you would normally import the private key or seed phrase into a wallet program. Protect paper wallets from physical harm.

WHAT IS MULTISIG?

The term "multi-signature" or "multisig" refers to a feature in Bitcoin that enables multiple signatures to be needed to approve a transaction. By demanding consent from many parties before a transaction can be carried out, it adds an extra degree of protection and control. Multiple public keys are used to create multisig addresses, which need a certain number of associated private keys to sign a transaction. To withdraw Bitcoin from a 2-of-3 multisig address, you would need two of the three private keys. This technique can be helpful in scenarios when many parties need to jointly manage and control Bitcoin funds, such as in business partnerships, escrow services, or shared wallets. Multiple signatures are necessary to limit the chance of a single point of failure or unauthorized access.

WHAT IS A BITCOIN FULL NODE?

A computer application known as a Bitcoin full node keeps a complete copy of the Bitcoin blockchain and takes part in the network's transaction and block validation and propagation. The decentralization and security of the Bitcoin network are greatly aided by full nodes. They ensure that the network works in accordance with the consensus rules by verifying all of the rules and transactions of the Bitcoin protocol. Incoming transactions and blocks are verified by full nodes, which also record the complete Bitcoin transaction history. Additionally, they broadcast fresh transactions and blocks to other nodes, assisting in their distribution across the network. People may help the Bitcoin network remain robust and censorship-resistant by operating a full node. Anyone can manage full nodes.

WHAT IS A BLOCK EXPLORER?

A web-based tool or program known as a Bitcoin block explorer is used to explore and view data related to Bitcoin transactions, blocks, and addresses. It offers a simple user interface for searching, examining, and following transactions on the Bitcoin blockchain. Users can get information about specific transactions or blocks using a block explorer by entering a transaction ID, block height, or Bitcoin address. The transaction value, sender and recipient addresses, timestamps, and the number of confirmations are all displayed by block explorers. Additionally, they display the blockchain's most recent block, the total number of transactions, and network statistics. Block explorers are useful tools for ensuring Bitcoin transactions are transparent, accountable, and in good standing. They are frequently used to learn more about Bitcoin activity.

WHAT IS A BITCOIN MINING POOL?

A Bitcoin mining pool is a cooperative group of miners who pool their computing power to jointly mine new blocks and split the profits that emerge from their efforts. Individual miners can pool their resources through mining pools, improving their odds of successfully mining a block and earning the corresponding block reward. According to the hash power each miner supplied, the reward is divided among the participating miners when a mining pool successfully mines a block. Individual miners who might not have enough computing power to mine blocks separately may benefit from joining a mining pool. By sharing the mining rewards together, it offers a more steady income source. In order to cover their costs, mining pools often take a modest cut of the profits.

WHAT IS A BITCOIN MINING RIG?

The specialized hardware set-up used for mining Bitcoin is referred to as a "mining rig." Mining rigs are specialized devices created to effectively carry out the intricate mathematical computations required to mine new blocks since mining demands a substantial amount of computational power. Due to their increased efficiency and performance, these rigs are outfitted with specialized mining hardware known as application-specific integrated circuits (ASICs), which have grown in popularity for Bitcoin mining. Multiple ASICs or GPUs are frequently connected to a motherboard and power supply to make up mining rigs. Mining software is used to manage and improve the mining activities, and rigs are normally set up in cold settings to prevent overheating. One setup that has increased in popularity is an immersion cooled Bitcoin mining rig.

CAN I MINE BITCOIN WITH MY PERSONAL COMPUTER?

It is technically possible to mine Bitcoin using a home computer but doing so is rarely practical. In order to have a reasonable chance of mining blocks and receiving rewards, Bitcoin mining has developed into a fiercely competitive sector that calls for tremendous computer power and specialized technology. Personal computers lack the hashing capacity to compete with the specialist mining rigs and mining farms that predominate the network, especially those with ordinary CPUs or consumer-grade GPUs. Additionally, mining is economically impractical because the expenses of electricity are frequently higher than the possible profits.

WHAT IS BITCOIN'S ENERGY CONSUMPTION?

Due to the energy-intensive nature of the mining process, the energy usage of Bitcoin has been a subject of dispute. Mining needs a lot of computational power, which translates into a lot of energy use. Being dependent on variables like the total network hash rate, the type of mining hardware being used, and the energy effectiveness of mining activities, the precise energy consumption of the Bitcoin network is difficult to quantify. The advantages that Bitcoin offers, like as a decentralized and censorship-resistant monetary system, should be weighed against its energy usage.

WHAT IS BITCOIN'S ENVIRONMENTAL IMPACT?

The main causes of Bitcoin's environmental effect are its energy requirements and the energy sources employed in mining. Concerns about mining's carbon impact have been highlighted due to its energy-intensive nature and some regions' dependency on fossil fuel-based electricity. Coal and other non-renewable energy-based mining practices contribute to greenhouse gas emissions and environmental damage. It's crucial to remember that the type of energy utilized for Bitcoin mining differs depending on where you are. The environmental impact of Bitcoin mining is considerably lessened in areas where renewable energy sources, like hydroelectricity or solar power, are widely available and reasonably priced. Greener mining techniques, such as the use of renewable energy, are being promoted as

the industry develops and awareness of sustainability rises.

WHAT ARE BITCOIN TRANSACTION FEES?

Users must pay miners for the privilege of having their transactions added to the blockchain through the use of Bitcoin transaction fees. You can choose to include a transaction fee when sending a Bitcoin transaction, which encourages miners to give it priority and include it in the upcoming block. For their computational work in protecting and validating the Bitcoin network, miners are rewarded with transaction fees. The dynamics of the network's supply and demand dictate the fees. Users might have to pay additional fees to guarantee that their transactions are processed swiftly when the network is overloaded with a large number of transactions. Lower fees may be employed when network activity is minimal. Various elements, including transaction size, network congestion, and transaction urgency, might affect transaction fees.

Wallets typically give an estimate of the suggested cost based on the state of the network. It's crucial to understand that transaction fees are distinct from any exchange or brokerage charges related to purchasing or selling Bitcoin.

CAN BITCOIN BE USED FOR CHARITABLE DONATIONS?

Bitcoin can be used to make charitable contributions, and the nonprofit community has embraced it more and more. The benefits of using Bitcoin for charitable contributions include its potential for privacy and borderless nature. Many nonprofits and charity institutions now accept Bitcoin as a means of donation. To donate Bitcoin, you normally need to look for a charity that takes it and adhere to their detailed guidelines. Giving your Bitcoin address or utilizing a payment processor to process the donation may be required. Depending on your country, giving Bitcoin can have advantages including transparency, accountability, and possible tax advantages. Make sure the charity has a solid track record and uses donations in a responsible manner by doing some research on it. Also, consulting with a tax professional can help you understand the potential tax implications of donating Bitcoin.

CAN GOVERNMENTS BAN BITCOIN?

The likelihood that governments will outlaw Bitcoin differs by region. Others have adopted a more cautious stance or set limits, while some governments have embraced Bitcoin and implemented legislation to control its use. The global accessibility and decentralized nature of Bitcoin make a total ban on it technically difficult. Governments have the power to control and regulate Bitcoin-related activities including exchanges, enterprises, and tax requirements. While some nations have issued warnings or limited specific features of Bitcoin, others have passed laws to establish legal foundations for cryptocurrency activity. To ensure that Bitcoin is used lawfully and in accordance with local regulations, it is crucial to be aware of these criteria. Governments' stance on Bitcoin can also alter. Informational upkeep is essential. One things is certain, government can ban Bitcoin but they cannot kill Bitcoin.

CAN I SEND BITCOIN TO SOMEONE WITHOUT A BITCOIN WALLET?

No, it is not possible to transmit Bitcoin straight to a person without a Bitcoin wallet. The recipient needs to have a Bitcoin wallet with a specific Bitcoin address in order to receive Bitcoin. The place where the money is sent is specified by the Bitcoin address. The recipient won't have an address to receive the money if they don't have a Bitcoin wallet. Before you can send someone Bitcoin, they must first create a Bitcoin wallet and provide you their Bitcoin address. Software wallets, hardware wallets, and web-based wallets are just a few of the different kinds of Bitcoin wallets that are accessible. Every style of wallet offers unique benefits and security considerations.

WHAT IS BITCOIN'S RELATIONSHIP WITH THE DARK WEB?

The term "dark web" refers to a region of the internet that search engines do not index and is frequently connected to illegal markets and activities. Some users on the dark web have become interested in Bitcoin because of its pseudonymous nature and possibility for anonymous transactions. The public blockchain, which offers a visible and immutable transaction history, stores Bitcoin transactions. Instances of people engaged in unlawful activity on the dark web have resulted in prosecutions. Law enforcement organizations have developed tools and strategies to track and detect illicit Bitcoin transactions. Silk Road was a dark web marketplace that facilitated the unlawful trading of numerous drugs, firearms, and other illegal items and services. Ross Ulbricht, the marketplace's founder, was found guilty and given many life sentences in jail before the marketplace was shut down by the FBI. Bitcoin served as the main form of payment on Silk Road. Buyers would deposit Bitcoin into an escrow account, which would then be given to the vendor once the transaction was complete. Bitcoin was the only form of payment accepted on Silk Road. The usage of Bitcoin on Silk Road increased public awareness of Bitcoin's potential application in illegal operations and raised questions about its connection to crime. The vast majority of Bitcoin transactions are for legal reasons, and businesses, people, and governments all around the world are utilizing the technology for a variety of legal purposes.

CAN BITCOIN BE USED FOR MONEY LAUNDERING?

Since it may help transport money across borders and offer some level of secrecy, Bitcoin can be used for money laundering. Bitcoin isn't built from the ground up to be used for money laundering, and the technology is transparent and traceable. The identities behind Bitcoin transactions are not directly connected to real-world identities due to the pseudonymous nature of the currency. Tracking and locating the parties engaged in a transaction may be difficult as a result of this feature. With the use of technology, regulatory organizations and law enforcement may examine blockchain data and spot questionable activity. Know Your Customer (KYC) and Anti-Money Laundering (AML) legislation, which demand customer identification and surveillance to stop illegal actions, are applicable to exchanges and other Bitcoin

service providers.

WHAT PLACES HAVE INTEGRATED BITCOIN INTO THEIR GOVERNMENTS?

Many places have expressed interest in implementing policies that are conducive to Bitcoin or integrating it into their governments. El Salvador became the first nation to accept Bitcoin as legal money in September 2021. The government put into effect the "Bitcoin Law," which permits the use of Bitcoin in regular transactions and mandates that companies accept it as payment. The government also is doing Bitcoin mining with volcanos. In short, El Salvador has become the primer choice and the darling of the Bitcoin community as they offer citizenship based on people who work in the Bitcoin industry. Second, the Ukrainian government has stated that it supports blockchain technology and Bitcoin. To establish a trustworthy and vibrant cryptocurrency industry, they have suggested laws to legalize and regulate cryptocurrencies. Thirdly, the regulatory climate for blockchain technology and cryptocurrencies is favorable in Switzerland. With multiple blockchain businesses and foundations functioning there, the Swiss government has taken efforts to foster innovation in the industry.

WHAT IS BITCOIN'S CORRELATION WITH TRADITIONAL FINANCIAL MARKETS?

The relationship between Bitcoin and conventional financial markets is complex and subject to change over time. At first, the value of Bitcoin was mostly determined by variables unique to the Bitcoin ecosystem as a whole, making it appear to be a distinct and uncorrelated asset class. During times of market stress or economic uncertainty, Bitcoin has occasionally shown a positive correlation with certain assets, such as equities or gold. This is due to the possibility that investors would view Bitcoin as a store of value or an inflation hedge comparable to gold. However, there have also been times when Bitcoin has not correlated at all with traditional markets, underlining its own features and potential benefits of diversification. It's best to think about

Bitcoin as a distinct, one-of-a-kind asset class.

WHAT ARE BITCOIN FUTURES?

Financial contracts known as "Bitcoin futures" enable investors to make predictions about the price of the digital currency without actually holding any of the underlying assets. These futures contracts bind the buyer to buy Bitcoin or the seller to sell Bitcoin on a predetermined future date at a predetermined price. With the use of Bitcoin futures, investors can take long (buying) or short (selling) positions, benefiting from both increasing and falling values. Exchanges trade futures contracts for Bitcoin. Instead of receiving Bitcoin in person, these contracts are settled in cash. The difference between the contract price and the market price of Bitcoin is settled in cash on the designated future date. Bitcoin futures give investors exposure to price changes without requiring them to keep or actively manage the underlying asset. They also come with dangers, such as market volatility and the possibility for large price movements, but they also provide liquidity and the ability to leverage bets.

CAN I BUY FRACTIONS OF A BITCOIN?

Yes, it is possible to purchase Bitcoin in small amounts. The lowest unit of Bitcoin, which can be divided into smaller units called "satoshis," is named after the mysterious person who invented Bitcoin, Satoshi Nakamoto. The value of one Bitcoin is one hundred million satoshis (0.00000001 BTC). The flexibility of owning and conducting transactions with lesser quantities of Bitcoin is made possible by its divisibility. Depending on your budget and investing objectives, you can buy a full Bitcoin or a portion of one. Instead of having to buy a whole Bitcoin, many exchanges allow customers to invest with smaller amounts of money by allowing them to acquire fractions of a Bitcoin. This makes Bitcoin available and permits users with different financial capacities to participate.

WHAT IS A 51% ATTACK?

When one miner or a group of miners control more than 50% of the network's total computing power, the situation is referred to as a 51% attack. They may be able to influence transactions or disrupt the network thanks to their majority control. Due to the widespread distribution of the Bitcoin network's miners, it is becoming more and more challenging for a single entity to gather such a sizable share of the network's computational power. A 51% attack is less likely to go undetected or unopposed since the Bitcoin community actively monitors and responds to any possible threats. The powerful proof-of-work consensus algorithm and the decentralized architecture of the Bitcoin network make it resistant to intrusions and ensure the network's integrity and security.

WHAT IS A DUST ATTACK?

Sending extremely little sums of Bitcoin (typically cent-sized amounts) to numerous addresses constitutes a dust assault. Such an attack aims to link addresses and perhaps de-anonymize people rather than to steal money. An attacker can examine subsequent transactions by sending dust to multiple addresses in an effort to look for patterns or connections between them. This attack approach takes use of the openness of the Bitcoin blockchain, where addresses and transactions are available to everyone. Use caution and stay away from dealing with dust transactions in order to defend against dust attacks. Avoiding accumulating or spending large amounts of dust can reduce the chance of being targeted or recognized by this attack vector.

CAN BITCOIN BE USED FOR MICRO PAYMENTS?

Bitcoin enables transactions with even its smallest units, known as satoshis. There are a few things to keep in mind, though. First off, during times of network congestion, Bitcoin transaction fees can become rather high. Micropayments may become less cost-effective as a result, as the fees may exceed the cost of the actual transaction. Second, the time it takes for a Bitcoin transaction to be confirmed can differ, with low-value transactions potentially taking longer. When it comes to instant micropayments, this might have an impact on the user experience. Layer 2 alternatives, such as the Lightning Network, have arisen to address these problems. Faster and less expensive off-chain micropayments are made possible via the Lightning Network, which functions as a second layer on top of the

Bitcoin blockchain.

WHAT IS A BITCOIN QR CODE?

The easiest way to share a Bitcoin address is with a QR code. The alphanumeric representation of a Bitcoin address is encoded in a two-dimensional barcode. This QR code can be scanned by Bitcoin wallet software to have the recipient's address prefilled when making a transaction. The manual entry of lengthy Bitcoin addresses is simplified with QR codes, lowering the chance of mistakes and enhancing the user experience when sending or receiving Bitcoin. A Bitcoin enthusiast by the name of Andreas Antonopoulos is largely credited with creating and popularizing the Bitcoin QR code. Antonopoulos popularized the use of QR codes as a technique to make Bitcoin transactions simpler and more effective, especially for mobile devices, in his book "Mastering Bitcoin: Unlocking Digital Cryptocurrencies". Antonopoulos realized that it was possible to utilize QR codes to store payment or wallet information for Bitcoin, allowing users to rapidly scan the code with their smartphones rather than typing out lengthy alphanumeric strings. This development made sending and receiving Bitcoin payments considerably easier and contributed to the widespread acceptance of Bitcoin as a more approachable and user-friendly digital currency. In Bitcoin, QR codes are frequently employed.

WHAT ARE DLCS?

Discreet Log Contracts (DLCs) are a kind of smart contract that enable conditional payments on Bitcoin while concealing the contract's specifics from outside observers. DLC transactions are made to look identical to other Bitcoin transactions because to this discretion, boosting both privacy and effectiveness. In DLCs, participants agree to swap money depending on how an oracle predicts an event will turn out. This technique can be used by developing an oracle that agrees to publish the value of a data stream at a specified time and then expose that value and signature when it is time. As an alternative, lit, a Lightning Network implementation, can be used to run DLCs. Many within the Bitcoin community are anticipating the mass adoption of Bitcoin as well as DLCs for the sake of fiscal sovereignty.

WHAT IS THE LIGHTNING NETWORK?

On top of the Bitcoin blockchain, the Lightning Network is a second-layer protocol designed to alleviate Bitcoin's scalability issues. By enabling participants to establish off-chain payment channels, it is intended to facilitate quicker and less expensive transactions. These payment methods enable transactions between users without necessitating the recording of every transaction on the Bitcoin blockchain. Transactions are instead carried out off-chain and can be settled immediately. Smart contracts and multi-signature wallets are used by the Lightning Network to promote participant security and trust. Micropayments are made possible, and the Bitcoin network's capacity for transactions can be greatly increased. The Lightning Network presents a potential solution to the scalability issue by lowering the

burden on the primary blockchain, also referred to as Layer 1,

and enabling swift transactions.

CAN I USE BITCOIN FOR REOCCURRING PAYMENTS?

There are numerous platforms and third-party services that permit recurring Bitcoin payments. By acting as a middleman between the payer and the payee, these services enable the automation of typical Bitcoin payments. Users have the option of setting up recurring payment schedules and giving these providers permission to make payments on their behalf.

WHAT IS TAPROOT?

In January 2018, Bitcoin Core developer Gregory Maxwell first suggested using Taproot. On June 12th, 2021, the majority of Bitcoin miners came to an agreement for the idea to be put into action. On November 14, 2021, Taproot becomes active at block 709,632. The advantages of Taproot for efficiency and privacy go hand in hand. Less transaction data is available to anyone examining the publicly accessible ledger for Bitcoin by committing fewer transactions to the blockchain. Less data is committed, allowing for more transactions to fit into each block, which should result in lower fees and higher transaction throughput. Additionally, Taproot paves the way for more Bitcoin smart contract flexibility. While it has long been simple to write and enforce complicated smart contracts on other blockchain networks, Taproot is expected to lay the technical groundwork for DeFI (Decentralized Finance) to quickly take off on the

Bitcoin network. With the use of Taproot, Bitcoin would be able to host enormous smart contracts with thousands of signatures, all while keeping the anonymity of all parties and the size of a single-signature transaction.

WHY SHOULD I
LOOK AT BITCOIN
L2 BLOCKCHAINS?

Layer 2 blockchains for Bitcoin provide scalability, quicker confirmation times, cheaper transaction costs, improved privacy, and new avenues for innovation and decentralized applications. A fuller knowledge of the developing Bitcoin ecosystem and the opportunities it offers consumers and developers can be gained by investigating these solutions. Online in 2010, Satoshi Nakamoto expressed his willingness to support a blockchain and network that pooled Bitcoin's processing power. Currently, there is what I would call tribal infighting among Bitcoin zealots who are anti L2 because they are exclusively pro Bircoin. To this, I say that Satoshi believed it to be ideal to have a seperate network that uses the CPU power of Bitcoin for scalability. If it scales Bitcoin and does not disturb

Bitcoin, I believe it is good for Bitcoin. It's also worth noting that innovation on Bitcoin started in 2010 with Namecoin, 2012 with Colored Bitcoin, 2014 with Counterparty, 2016 with Rare Pepes, and more.

WHAT IS STACKS?

Stacks is a Bitcoin L2 blockchain that enables smart contracts and decentralized applications on Bitcoin. Stacks, formerly known as Blockstack, was created by Muneeb Ali and Ryan Shea. Blockstack is known for being the first token distributed through the first-ever U.S. Securities and Exchange Commission (SEC) qualified token offering in 2019. After the launch of the Stacks Blockchain 2.0, Stacks was deemed decentralized. With increased capability and programmability, the Bitcoin blockchain maintains its security and decentralization. The Clarity programming language used by Stacks is safeguarded by the whole Bitcoin hashrate and is designed for secure, decidable contracts. This allows for more functionality and programmability while maintaining the security and decentralization of the Bitcoin blockchain. A decentralized network of independent businesses, programmers, and community members backs

Stacks. The Stacks community is motivated to build a user-owned internet using Bitcoin as the settlement layer.

WHAT IS ROOTSTOCK?

Rootstock is a sidechain that aims to provide smart contract capabilities on top of the Bitcoin blockchain. Rootstock supports Ethereum Virtual Machine (EVM), allowing developers to use Ethereum-style smart contracts on the Bitcoin network. Rootstock was birthed from the efforts of Sergio Lerner, Diego Gutierrez Zaldivar, and Nick Szabo in 2015. Rootstock is protected by Bitcoin's proof-of-work algorithm, which leverages the security of the main Bitcoin blockchain. Rootstock's native token is RBTC (Smart Bitcoin), and it allows users can utilize their BTC to interact with decentralized applications (dApps) established on the Rootstock network. Rootstock is classified as a Layer 2 solution for Bitcoin since it expands the network's capabilities without changing the basic Bitcoin protocol. Rootstock's major use cases include allowing DeFi (decentralized financial) applications, cross-chain interoperability, and improving the overall scalability and capabilities of the Bitcoin ecosystem.

WHAT ARE .BTC NAMES?

The .btc extension for human-readable names connected to Bitcoin addresses was created by Stacks, who also created the Bitcoin Name System (BNS). The infrastructure and tools needed to register and manage .btc names are provided by the Stacks ecosystem. Users can reserve their desired .btc names and link them to their Bitcoin addresses using Stacks' BNS protocol. To guarantee the authenticity and ownership of the registered names, Stacks' BNS makes use of the security and immutability of the Bitcoin blockchain.

WHAT IS A BITCOIN VANITY ADDRESS?

A personalized Bitcoin address known as a "vanity address" is one that has been designed with a certain word, name, or pattern in mind. By continuously generating public-private key pairs until they find one that matches their desired pattern, users create vanity addresses. Vanity addresses can be used for branding, customization, or advertising, but they don't offer any more security or functionality than standard Bitcoin addresses.

WHAT ARE ORDINALS?

Digital assets inscribed on a satoshi, the smallest unit of value in a Bitcoin, are known as ordinal inscriptions. These are akin to NFTs. Satoshi Nakamoto's token, the satoshi, can now be inscribed on owing to the Taproot upgrade that was introduced to the Bitcoin network on November 14, 2021. They have taken the world by storm and have brought attention, developers, revenue for Bitcoin miners, and renewed innovation to Bitcoin.

WHAT ARE PSBT?

Partly Signed Bitcoin Transactions (PSBTs) are a standard way for many participants to sign a transaction cooperatively inside the Bitcoin ecosystem. PSBTs make use of a particular data format that makes it easier for wallets and other tools to transmit details about a Bitcoin transaction and the related necessary signatures. This format facilitates cooperative transaction design, transfer, and signing, making it especially helpful for transactions involving multisig addresses. The use of PSBTs encourages compatibility between other wallet programs, making it possible to create transactions in a watch-only wallet and then export them to another wallet for signing and broadcasting. Several programs can now sign the transaction, whether for multisig wallets or various inputs, thanks to PSBTs. Before being transformed into a complete transaction, several PSBTs may be concatenated into a single completely signed PSBT. In Bitcoin Improvement Proposals (BIPs) like BIP174 and BIP370, the specifications and rules for PSBTs are detailed.

WHAT ARE BRC-20 TOKENS?

BRC-20, or Bitcoin Request for Comment 20, was launched in March 2023 by an unidentified developer going by the name Domo. It is based on the Ethereum protocol ERC-20 (Ethereum Request for Comment 20). With a few significant exceptions, such as the absence of smart contracts, BRC-20s are essentially Bitcoin's equivalent of ERC-20s. The Bitcoin network was not initially designed to support such use cases so it has led some to criticize the use of BRC-20 tokens on Bitcoin, arguing that they may conflict with the network's original purpose and add unnecessary complexity. Regardless, BRC-20 tokens presently remain a popular way for developers and startups to build on top of the Bitcoin blockchain and leverage its network effects.

WHAT ARE SRC-20 TOKENS?

An alternative approach to incorporating picture data into Bitcoin is the emerging STAMPS (Secure Tradeable Art Maintained Securely) protocol. "Bitcoin Stamps" hold picture data directly within spendable transaction outputs as opposed to putting it in prunable transaction witness data. Another benefit of stamps is that they are "semi-fungible"; they can be printed as "1 of 1" or "1 of many" digital assets.

CAN I DEVELOP
ON BITCOIN?

To create Bitcoin applications, you can utilize a variety of programming languages. A variety of Bitcoin libraries, including Python Bitcoinlib and Pybitcointools, are available in the flexible and user-friendly language Python for dealing with the Bitcoin network. A common language for web development is JavaScript, which may be used with frameworks like BitcoinJS to build web-based Bitcoin apps. Go is a fast, statically typed programming language that offers tools like btcsuite for creating Bitcoin applications that are performance-focused. The main programming language used to create the Bitcoin Core program, which runs the Bitcoin network, is C++. Low-level control is possible, as is performance improvement. A systems programming language called Rust is well renowned for emphasizing concurrency and safety. The tools required for Bitcoin development are offered by Rust libraries like Rust Bitcoin. New technological advancements are being made on Stacks, such as sBTC and the Nakamoto release, and Clarity is a language used to scale Bitcoin over the Stacks blockchain through smart contracts. Finally, when it comes to creating with Bitcoin, Miniscript is a fantastic place to start for those searching for a native Layer 1 experience. Select a language that is compatible with your development expertise, your project's requirements, and the libraries and tooling that are offered for Bitcoin development in that language.

WHAT IS A BIP?

A Bitcoin Improvement Proposal (BIP) is a formal proposal for changes to the Bitcoin blockchain. BIPs cover a wide range of potential modifications, such as adjustments to the consensus layer, community standards, or development procedure of Bitcoin. These ideas offer a standardized framework for suggesting and executing changes to the protocol, which is essential for the Bitcoin ecosystem. The BIP framework makes sure that all system changes are subjected to meticulous analysis, extensive public scrutiny, and community consensus-building. The BIP goes through several processes, such as drafting, proposing, and approval, before being implemented by the nodes responsible for maintaining the Bitcoin network. As the community engages in considerable deliberation and iteration, the implementation process for BIPs can be time-consuming, frequently lasting years.

HOW CAN ANYONE CONTRIBUTE TO BITCOIN?

Create educational content, join in online forums and social media platforms, or write blog articles to convey knowledge about the advantages and potential of Bitcoin. You can contribute to Bitcoin's open-source development if you have programming abilities. The source code for Bitcoin is openly accessible on websites like GitHub, where programmers can take part in enhancing the code, suggesting and implementing new features, or doing security audits. Encourage your jurisdiction to adopt Bitcoin-friendly laws and regulations. Engage with local representatives, go to cryptocurrency conferences or events, and take part in debates regarding the value of blockchain technology and financial innovation. Shop at establishments that take Bitcoin. By actively using Bitcoin for your purchases, you promote its use as a money and aid in its adoption. Join groups on social media or online forums that are focused on Bitcoin. Participate in debates, impart knowledge, and assist in addressing inquiries from newcomers. This helps to create a lively and encouraging community. Participate in academic and research projects centered on blockchain and Bitcoin, cooperate with universities and research organizations to publish papers, carry out investigations, or otherwise increase our understanding of cryptocurrencies. No matter how big or small, contributions can have an impact on Bitcoin's development and uptake. Find a niche that fits your interests and skills, and actively engage in the Bitcoin community to support its growth.

WHAT IS BITCOIN PIZZA DAY?

May 22, 2010 is known as Bitcoin Pizza Day. On that day, Laszlo Hanyecz, an early adopter of Bitcoin, paid 10,000 Bitcoins, or less than a cent per coin, for two pizzas from Papa John's Pizza. This was the first transaction in which Bitcoin was used as money in the real world, and it has since come to represent this moment in Bitcoin and cryptocurrency history. The cryptocurrency community commemorates Bitcoin Pizza Day today to highlight how far Bitcoin and blockchain technology have come since their inception and how they are still developing and having an impact on other industries in addition to finance.

HOW HIGH OR LOW WILL BITCOIN GO?

Many suggest that 1 Bitcoin will be, at least, the equivalent of $1,000,000. The case for Bitcoin attaining a price of $1 million per Bitcoin is based on its rarity, rising use, and acceptance as a major worldwide store of wealth. There are roughly 18.7 million Bitcoins in circulation out of a maximum 21 million available. Bitcoin's price may increase as more individuals and organizations use it as a form of inflation protection and a store of value. The usage of Bitcoin exchanges, which let users buy and sell Bitcoin at market rates, is a common technique to keep track of price changes in the virtual currency. Numerous websites and smartphone programs, such CoinGecko and TradingView, offer real-time price tracking and analysis of Bitcoin. Additionally, social media platforms and financial news sources regularly cover changes and trends in the price of Bitcoin.

As far as how low Bitcoin can go, it's important to remember that it was less than a fraction of a penny at one point. It's difficult to quantify how low a price can go. There are bear markets where everything is red and there are bull markets where everything is green. Volatility comes in waves but Bitcoin has been on the steady rise for 14 years. No one truly knows how high or low Bitcoin will go so it is best to remember that this is not financial advice and to do your own research as well as do what is fiscally best for you.

DOES BITCOIN CONTRIBUTE TO FINANCIAL INCLUSION?

By giving underbanked or unbanked people access to financial services, Bitcoin has the potential to support financial inclusion. Anyone with an internet connection and a suitable gadget can access Bitcoin. This implies that people can still participate in the global financial system and store value digitally even if they do not have access to traditional banking services. Low-cost and quick cross-border transactions are made possible by Bitcoin's decentralized and borderless nature. It can be a more effective way for people to transfer and receive money across borders than the pricey and time-consuming traditional remittance services. Bitcoin can act as a decentralized, local currency-independent store of value in areas experiencing political or economic unrest. It provides people with a way to secure their wealth from inflation and governmental capital controls. Bitcoin gives users direct ownership and control over their money, putting them in charge of their own finances. It does away with the necessity for middlemen like banks and gives people the freedom to handle their financial affairs on their own.

WHAT IS HYPER BITCOINIZATION?

Hyperbitcoinization is a hypothetical situation in which Bitcoin replaces fiat currencies and other conventional forms of money as the main form of money on the planet. In a hyperbitcoinization scenario, Bitcoin would be extensively used for the majority of financial transactions as well as as a unit of account and a store of value. One potential is that due to inflation, political unpredictability, or geopolitical concerns, people and institutions may start to lose faith in fiat currencies and turn to Bitcoin as a more reliable and safe alternative. Another possibility is that advancements in technology will make Bitcoin more useful and approachable for regular transactions.

HOW CAN BITCOIN BE A TOOL FOR GROUP ECONOMICS?

Bitcoin can be used for group economics since it allows people or organizations to transact without the use of middlemen, which can support and grow regional economies. By eliminating the additional fees associated with traditional financial services, Bitcoin can be used to facilitate local transactions between people or organizations in a specific neighborhood. By offering a more affordable method of receiving payments without the need for intermediaries and fees, Bitcoin can be a liberating tool for small businesses. Bitcoin makes it possible for communities to conduct autonomous fund-raising efforts for a variety of causes without depending on centralized organizations. The restricted supply and decentralized features of Bitcoin make it a potentially alluring choice for people or organizations looking to safeguard themselves against inflation or currency devaluation.

WHAT IS A BITCOIN CIRCULAR ECONOMY?

A Bitcoin circular economy (BCE) is a system of exchange and accounting that operates on the fundamental idea of using Bitcoin. Circular economies use Bitcoin to exchange value for goods and services rather than requesting permission to transact or using custodial rails with independent financial institutions that work in conjunction with State authorities. The term "circular" refers to the way that Bitcoin is recycled within the ecosystem to pay for products and services rather than being exchanged for fiat money to complete a transaction. It's a frequent misconception that a circular economy needs scale to be successful; yet, these economies must start small. Since it is an opt-in economy, one where many people are choosing to participate, the Bitcoin economy is not comparable to that of a nation-state and neither does it compel anybody to conform. The Surfer Kids

Non-Profit has a subsidiary called Bitcoin Ekasi. Founded in 2010, The Surfer Kids is a legally recognized nonprofit organization or NPO. An underprivileged South African township is the focus of the NPO, which is entirely funded by donations. Bitcoin was incorporated into the Surfer Kids project. The informal community of Bitcoin Ekasi in South Africa is situated in Cape Town, more specifically in Mossel Bay. Many BCEs are launching.

WILL BITCOIN BE A WORLD RESERVE ASSET?

Bitcoin's finite quantity, decentralized network, and network effect give it the potential to become a global reserve asset. Bitcoin has a deflationary and scarcity-based monetary policy because its total supply is limited to 21 million coins. As a result, Bitcoin is immune to the inflationary pressures that threaten to depreciate the value of conventional fiat currencies. Transactions may be made directly thanks to Bitcoin's decentralized network, which also serves as a decentralized store of value and an inflation hedge. The growing network effect of Bitcoin indicates that as more individuals, organizations, and governments accept and use it, its value and usefulness will rise, making it a desirable asset for both individuals and governments to keep. Countries may use Bitcoin as a means of hedging against inflation and geopolitical risks while diversifying their reserves away from conventional fiat currencies. Numerous characteristics of Bitcoin make it appealing as a possible world reserve asset. In order for this to materialize, there are now considerable regulatory and adoption obstacles that must be overcome. I agree with many who think that either an official adoption of Bitcoin as a global reserve asset or a grassroots adoption in which the majority of people (99%) chose Bitcoin as the top global reserve asset is imminent. By 2030, I predict that Bitcoin will be widely known and used by people all across the world, both in terms of acceptance and utilization.

IS BITCOIN THE FUTURE OF MONEY?

Bitcoin pioneered the idea of decentralized digital currency and offered a substitute for conventional fiat currencies governed by central banks. Blockchain, the technology at the foundation of it, has sparked creativity and sparked the creation of a number of cryptocurrencies and decentralized applications. Some people think of Bitcoin as a potential store of wealth that might rival gold because of its scarcity and limited quantity. For anyone looking to protect their capital and hedging against inflation or economic concerns, its decentralized structure and resistance to censorship make it appealing. Bitcoin has the potential to increase financial inclusion by giving underbanked or unbanked people access to financial services. People in areas with no access to traditional banking can participate in the financial system thanks to its borderless nature and capacity to facilitate inexpensive cross-border transactions. Blockchain technology, smart contracts, and decentralized finance (DeFi) applications have all advanced as a result of the rise of Bitcoin. These developments have the potential to transform a number of industries and upend established financial institutions. Governments and financial institutions throughout the world are scrutinizing and challenging the regulatory framework for Bitcoin and other cryptocurrencies. The adoption and general acceptability of Bitcoin as a form of money can be greatly impacted by the regulatory environment. It's crucial to remember that obstacles to Bitcoin's widespread acceptance as a mainstream currency include its present market price volatility and the changing regulatory landscape. Technology breakthroughs, legislative changes, public acceptance, and the coordinated activities of individuals and institutions in reshaping the financial landscape will ultimately determine the direction of money. If you ask me, I think Bitcoin is the currency of the future. Period.

AFTERWORD

This book addressed a wide range of concerns that readers frequently have regarding Bitcoin. We've looked at what a decentralized digital currency is, how blockchain technology works, and how it may be used as a store of value, a payment mechanism, and an investment. We've looked at mining, security precautions, and the regulatory environment related to this ground-breaking asset. We've discussed its potential for international remittances, potential environmental impact, and future scaling through Bitcoin Layer 2s like the Lightning Network, Rootstock, and Stacks. You are prepared to start your trip in the interesting world of Bitcoin if you have a firm grasp of these fundamental concepts. Remember to stay informed, use caution, and accept that Bitcoin is constantly evolving as you continue your exploration. This is a lifelong journey and you are equipped with the knowledge you need to continue learning and

even partake in the future of finance and applications, Bitcoin.

ABOUT THE AUTHOR

Christopher Perceptions

Christopher Perceptions stands at the intersection of AI, Bitcoin, and NoCode innovation, enabling Bitcoin to transcend its role as a mere currency to become the bedrock of creative development and equitable digital strategies. For nearly a decade, Christopher has been a driving force in shaping the future of decentralized technology.

As the CEO of NoCodeClarity, Christopher's vision is centered on mass adoption. His approach democratizes Bitcoin and Layer 2 development, making it accessible and comprehensible to both enthusiasts and entrepreneurs.

Author of the bestselling book "The Secrets of Satoshi: Understanding Bitcoin," Christopher has made a global impact through education. He has enlightened thousands and has been a featured speaker at premier conferences like Bitcoin Unleashed.

Through his weekly newsletter, "Christopher's Perceptions," he offers a wealth of knowledge, making complex topics in Bitcoin, Web3, and NoCode accessible to a global audience. Christopher is more than a leader; he's a coach, guiding entrepreneurs as they navigate and succeed in the ever-evolving Web3 domain.

A bridge between technical innovation and community engagement, Christopher reinforces his belief in Bitcoin as a system embodying fairness, innovation, and interdependence, paving the way for an impactful future.

www.ingramcontent.com/pod-product-compliance
Lightning Source LLC
Chambersburg PA
CBHW070928210326
41520CB00021B/6845